MW01005855

Kira and Lulu Visit the Planets

BY
Kat Davidson

To Noah - maybe the next book will be about your adventure!

Kira and Lulu Visit the Planets
By Kat Davidson
October 2018

Illustrations by Kat and Karen Davidson
Graphic Assistance by David Senkiw
Special Thanks: N. Schleuter, L. Cole, C. Demott

Artwork rendered in acrylic
Font Copyright Arial Rounded MT Bold

A project of Space Weather News LLC

ISBN: 978-0-57841-152-1

This book belongs to:

____

Our trip to the Sun was as fun as can be!

What other wonders are out there to see?

Our journey today will be just as fun... let's explore the planets that circle the Sun!

We will leave a quick note for Father and Mother, then prepare for a journey unlike any other.

Off to see all the planets we go!
It's time to start the planetary show!

3...2...1...

BLASTOFF!

Wave goodbye to the Moon as we fly
towards the Sun!
We're starting with

Mercury,

that is planet number one!

Mercury is rocky and
looks like the moon.

It is too hot near the Sun
so we better leave soon!

Up next is **Venus**, with yellow and brown skies.

Often called Earth's sister they are so close in size.

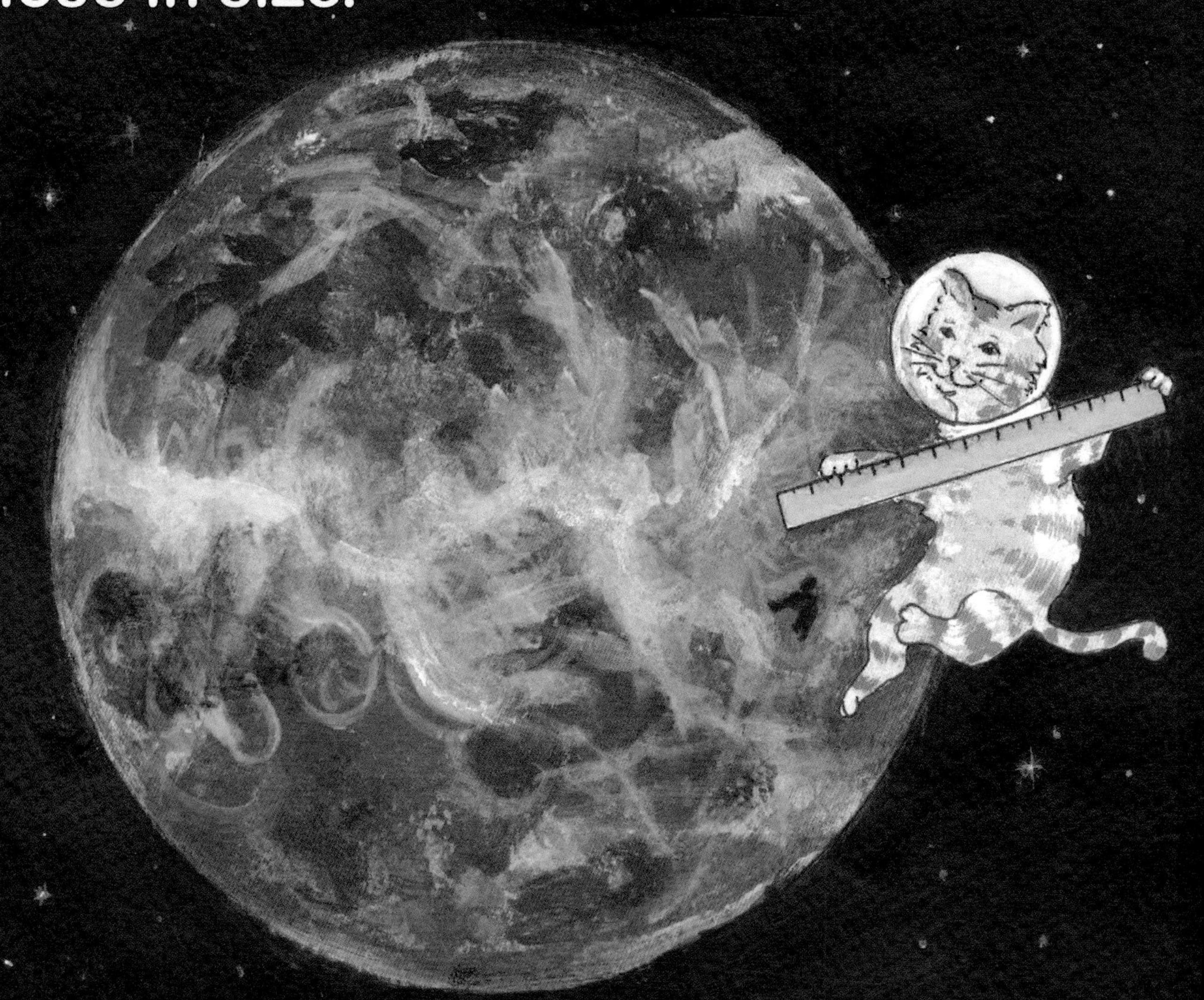

With water so blue and clouds so white, our home planet **Earth** is a beautiful sight!

As we fly past Earth and head to the stars, the next planet to see is a red one named Mars.

Mars is a red planet, the fourth from the Sun. Let's take a small break to do something fun!

There's ice all around here, and dirt many shades of red.

No plants or animals here so let's build castles instead!

Through the asteroid belt
Kira and Lulu fly,

dodging all of the rocks
as they sail on by.

Up next is Jupiter, it is the biggest planet around!

It's where great clouds, big storms and many moons can be found.

The Great Red Spot is the biggest storm of them all!

Bigger than three Earths together, don't you feel pretty small?

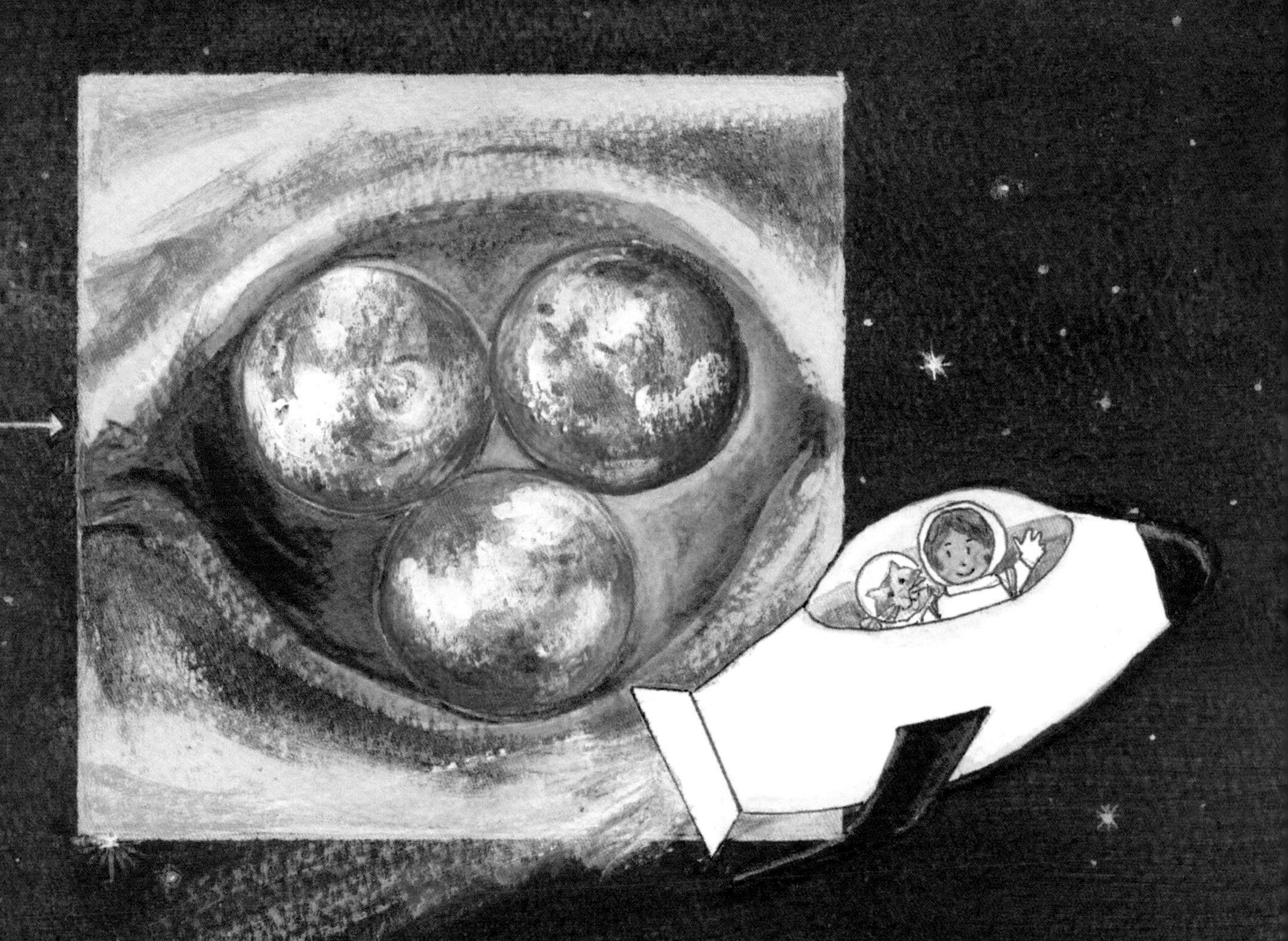

Let's speed towards **Saturn** and look at all of the rings.

So many colors and shapes- they are such pretty things!

Planet number seven is **Uranus**,
it looks tipped to the side
With smaller rings, strong winds,
stinky air far and wide.

We come upon **Neptune**, colored a beautiful blue. It's almost the end of the journey, so sad but it's true!

Pluto is the little one, and although it is small, it has some of the best colors and shapes of them all.

What is that light over there with a tail?

It's a bright COMET leaving an icy trail.

We follow the comet made of ice and rock- it seems to be speeding up, look at the clock!

We lost track of time, it has been so much fun, our trip to see the planets is almost done.

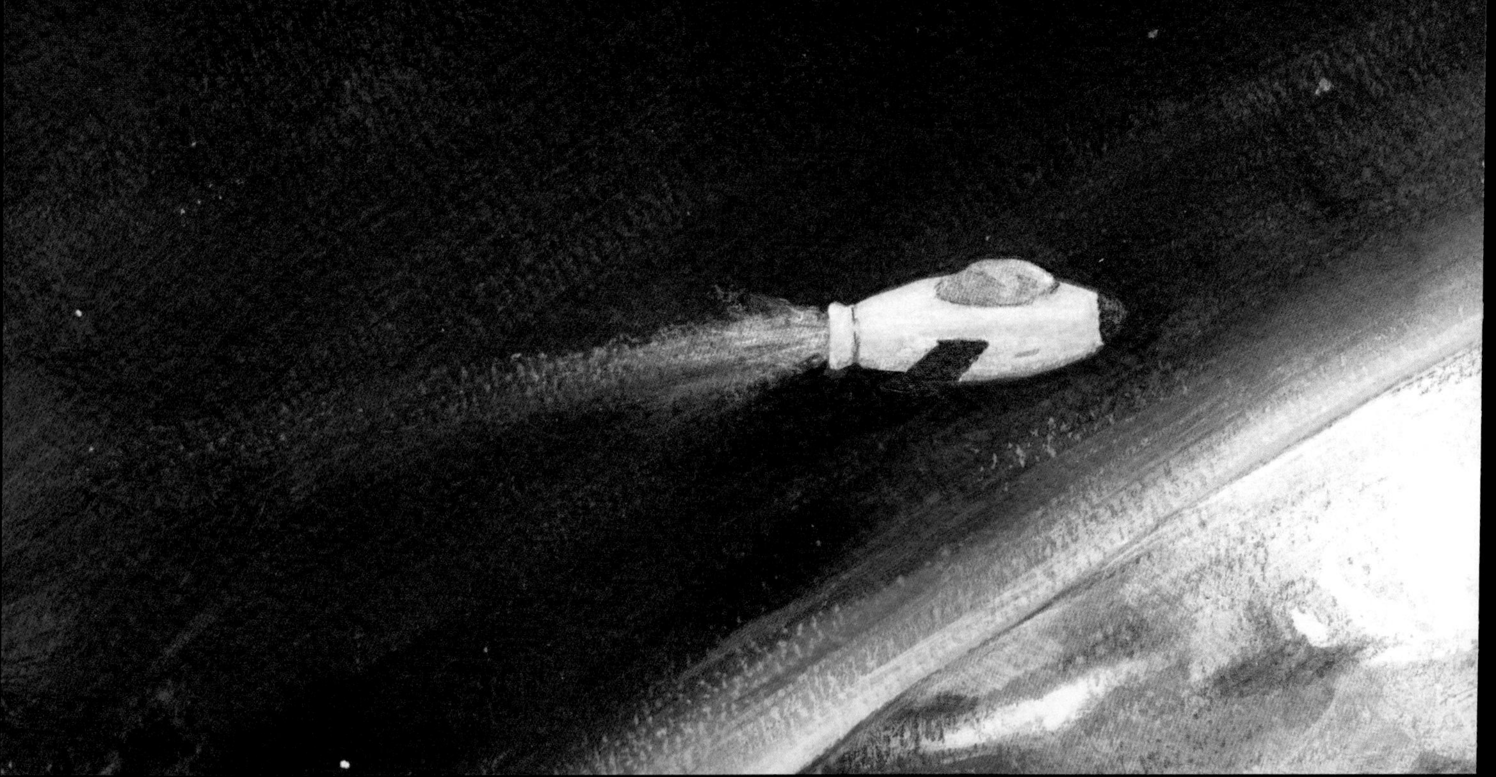

Before we go home let's remember the planets we've passed,
That belong in our solar system, we'll name them all from first to last:

VENUS

6

JUPITER

SATURN

3 EARTH

4 MARS

7 URANUS

8 NEPTUNE

9 PLUTO

GREAT JOB!

We arrive home in time to
see a comet fly by,
as we watch the
amazing colors
paint the sky.

"What a beautiful sight," say Father and Mother. "Maybe next time, you can bring little brother!"